contents

spacescent perfume bottle, 2000, detail

Joseph Rosa

design series 2

Yves Béhar fuseproject

SAN FRANCISCO MUSEUM OF MODERN ART

SFMOMA design series

This catalogue is the second in a series of exhibitions and accompanying publications intended to highlight the work of architects, industrial designers, and graphic designers at the forefront of their respective disciplines. This volume is published on the occasion of the exhibition *Yves Béhar fuseproject/design series 2,* organized by Joseph Rosa at the San Francisco Museum of Modern Art and on view from March 26 through October 3, 2004.

Previous books in the series: *ROY/design series 1*

Yves Béhar fuseproject/design series 2 is organized by the San Francisco Museum of Modern Art and has been generously supported by an anonymous donor and the LEF Foundation. Additional support has been provided by Pro Helvetia Arts Council of Switzerland, the Consulate General of Switzerland in San Francisco, Presence Switzerland, and swissnex.

Switzerland.

This book is set in Benton Gothic, available from the Font Bureau, Inc.

Library of Congress Cataloging-in-Publication Data

Rosa, Joseph.
  Yves Béhar fuseproject : design series 2 / Joseph Rosa.
    p. cm. — (SFMOMA design series ; 2)
Catalog of an exhibition at the San Francisco Museum of Modern Art.
Mar. 26–Oct. 3, 2004. Includes bibliographical references.
  ISBN 0-918471-71-0
  1. Béhar, Yves, 1967– —Exhibitions. 2. Design, Industrial—California—San Francisco—Exhibitions. I. Title: Fuseproject. II. Béhar, Yves, 1967– III. San Francisco Museum of Modern Art. IV. Title. V. Series.
  T180.S25 2004
  745.2'092—dc22

                    2003022413

Director of Publications: Chad Coerver
Managing Editor: Karen A. Levine
Art Director: Jennifer Sonderby
Designer: James Williams
Publications Assistant: Lindsey Westbrook
Printing: Celeste McMullin, the Printcess/Hemlock Printers
Printed and bound in Canada

Cover: Perfume09 bottle, 2002, detail
Back cover: Perfume09 bottle, 2002, rendering
Inside cover: conceptual rendering, 2003, exploration with Hussein Chalayan

All images courtesy Yves Béhar/fuseproject with photography credits as follows: cover: Eskil Tomozy/fuseproject; page 4: Alan Purcell; pages 8, 12 (center), 24, 25 (bottom), 30, 31, 33, 36 (left), 37, 40–41: Marcus Hanschen; page 12 (left), 26–27: Bert Spangmacher; pages 18–19: Nigel Cox; page 23: Hunter Freeman; page 35: Rick English; page 36 (center): Marc Serr; pages 38–39: fuseproject and Marcus Hanschen; page 43: Sven Wiederholt

Director's Foreword Neal Benezra

With this publication, the San Francisco Museum of Modern Art's Design Series turns its attention to the discipline of industrial design. In exploring the creative output of San Francisco–based Yves Béhar and his fuseproject studio, the series continues its distinctive focus on architects, industrial designers, and graphic designers whose work is revitalizing their respective fields.

Industrial design may be a younger discipline than architecture and graphic design, but in less than a century the profession has evolved both swiftly and dramatically. Once dominated by the teachings of the Bauhaus and the Ulm Design School, today the profession has split into two primary streams, one expanding on the traditional modernist approach to problem solving and functionality and the other forging new ground under the influence of architectural theory, contemporary art, and a variety of other fields. As curator Joseph Rosa points out in his catalogue essay, Béhar's work brilliantly represents this latter trend; the designer's innovative projects for Birkenstock, Herman Miller, HP, MINI, and Toshiba suggest that industrial design may be a conceptual pursuit—one that addresses the full arc of the user's experience, from product selection to transport, storage, and disposal. Today, as technologies continue to evolve at a blistering pace, projects by Béhar and his contemporaries not only provide crucial interfaces to the tools that serve us in our personal and professional lives, but also influence the ways in which we define and embellish our surroundings.

This exhibition and its accompanying catalogue would not have been possible without the close collaboration of Yves Béhar and his colleagues. Yves and Joseph Rosa join me in acknowledging fuseproject staff members Geoffrey Petrizzi, Johan Liden, Eskil Tomozy, Shawn Sinyork, Pichaya Puttorngul, Franz Schnaas, Youjin Nam, Catherine O'Connor, Ric Peralta, Amy Strickland, Reese Madrid, and Cameron Campbell for their invaluable contributions to the studio's activities. As always, we are grateful to Elaine McKeon, Helen Hilton Raiser, and the Board of Trustees for their ongoing advocacy of the Museum's design-related programming and to the Architecture and Design Accessions Committee and A+D Forum for their work on the department's behalf. The contributions of curatorial associate Ruth Keffer, the author of this catalogue's insightful project descriptions, were essential to the development of every aspect of the exhibition. The project also benefited from the participation of a number of other staff members, including Ruth Berson, deputy director, programs and collections; Olga Charyshyn, registrar, exhibitions; Alexander Cheves, museum preparator; Chad Coerver, director of publications; Steve Dye, exhibitions technical manager; Libby Garrison, interim director of communications; Karen Levine, managing editor, publications; Melissa Potter, assistant paper conservator; Kent Roberts, exhibitions design manager; Gregory Sandoval, manager of adult interpretive programs; Jennifer Sonderby, head of graphic design; Jill Sterrett, director of collections and conservation; Marcelene Trujillo, assistant exhibitions director; Lindsey Westbrook, publications assistant; and James Williams, designer. Finally, thanks are due to curatorial associate Darrin Alfred and administrative assistant Amy Ress for their support of all Architecture and Design Department activities.

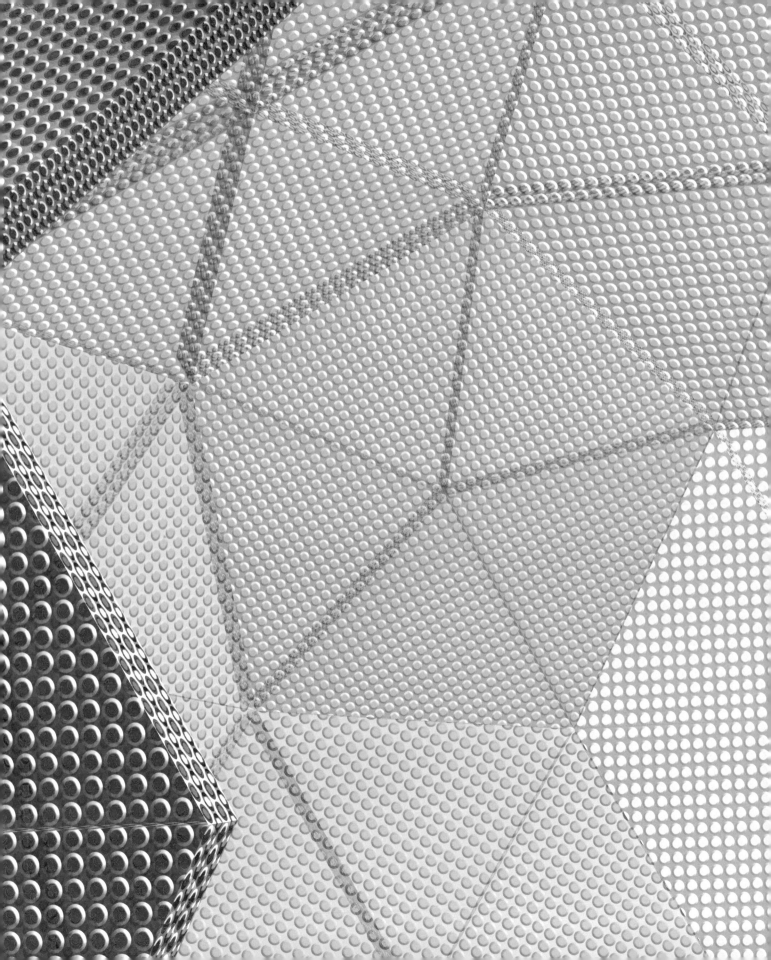

Brand Fusion: The Industrial Design of Yves Béhar   Joseph Rosa

At the age of fifteen Yves Béhar designed and produced his first hybrid product, a wind-powered board with two skis and a sail "that went frightfully fast" on frozen lakes.[1] This early invention points to a capacity for merging existing typologies to generate original and adaptive forms, an approach that continues to distinguish Béhar as one of the most innovative industrial designers working today. His ideology, which is founded on the construction of a conceptual narrative that guides interaction with the designed object, has allowed him to move seamlessly from classic industrial design to the creation of lifestyle objects and brand development over the course of his professional career. Béhar's method is emblematic of a new vision in industrial design that takes inspiration from the experience of the user.

Although the discipline has expanded radically since its inception, the pedagogy of industrial design remains rooted in a process of problem solving that ideally results in "good design"—a tradition grounded in the progressive teachings of the Bauhaus and the Ulm Design School. Today, though detached from the critical framework that once transformed the field, the problem-solving approach persists as a primary tool for many designers, who continue to apply it to products that were resolved aesthetically fifty years ago.[2] In the first half of the twentieth century, the profession embraced little-known designers such as Henry Dreyfuss, the creator of Bell Laboratories' iconic Model 300 telephone (1937), as well as flamboyant figures like Raymond Loewy, best known for his sleek, futuristic car designs for Studebaker, including the 1962 Avanti. The same holds true today, with lesser-known individuals working in large studios while brand-name designers such as Philippe Starck and Karim Rashid have become public personalities. Between these two extremes, however, a completely new breed of designer is emerging, moving beyond the age-old problems of functionality and employing a variety of methodologies that reflect current needs, contexts, materials, and technologies. Borrowing from other disciplines to build narratives around their products and to infuse them with unique character, Béhar and colleagues such as Ron Arad, Ronan and Erwan Bouroullec, Ross Lovegrove, and Marc Newson are redefining the nature of industrial design.

Béhar's ability to draw on diverse conditions to forge something new is perhaps the result of his multi cultural, cosmopolitan upbringing (he was born in Lausanne, Switzerland, in 1967 to a German mother and a Turkish father). Early on he realized that he was not interested in the traditional European industrial-design curriculum that had become the international standard for the discipline. Having grown up in a society that embraced the values of European modernism, Béhar felt it was essential to find new models that would help him see design as more than simply a problem-solving process.[3] He chose to study at the Art Center College of Design in Vevey, Switzerland, because its industrial-design pedagogy allowed access to influences from other artistic disciplines—a trend that was growing more prevalent in avant-garde design curricula. By the 1980s, design and architecture had begun to move away from the highly formal rhetoric of the Bauhaus, and Art Center's training for industrial designers reflected this tendency.

Opposite: Trium Erector, 2003, detail of rendering

Given his culturally diverse background, Béhar viewed it as "almost second nature to move to another country."[4] Having previously visited the East Coast of the United States with his parents, he went to the West Coast for the first time in 1989 to complete his degree at Art Center's main campus in Pasadena, California. In fall 1991, after serving a brief apprenticeship at Steelcase in Grand Rapids, Michigan, and earning his BS in industrial design, he decided to move to San Francisco.

The young industrial designer was attracted to Northern California because of its dynamic design culture; thanks to studios such as IDEO, frog design, and Lunar Design, "there was an enthusiasm for designing the machines of tomorrow."[5] Shortly after graduating, Béhar was hired by the San Francisco–based Burdick Group, where he worked for approximately nine months in 1992. He then moved to Lunar (1993–95) and frog (1995–98). In the summer of 1999 Béhar established his own studio, fuseproject, in the South Park neighborhood, the epicenter of San Francisco's dot-com boom (and subsequent bust).

Béhar credits his early years at Lunar and frog with "giving him the freedom to create and to manage significant clients."[6] But it was not until he went out on his own that he was able to position himself as a different type of industrial designer, making significant contributions to the discipline. Béhar's methodology results in designs that address the user's full experience of the object, primarily by constructing a narrative around it. This approach, which acknowledges conditions other than functional aspects, is more aligned with the experimental thinking that has transformed recent architectural pedagogy and critical practice.[7] For Béhar, many "industrial designers were just stuffing hard drives, creating features and products without considering the experience of using them."[8] It is precisely this lack of attention to a product's context that the fuseproject design process seeks to redress.

Imagining a narrative for each project enables Béhar to think about the consumer's intended interaction with the object, to identify a particular feeling associated with using the product, and to focus on that impression as the design evolves. In a process similar to storyboarding a film, fuseproject staff members brainstorm, share ideas, and then write a scenario for the product. Scripting helps Béhar answer important questions, including how to achieve the idea behind the concept and how to communicate it. Only after this process does he enter the drawing phase, which responds directly to these questions. In the final stage, fuseproject uses computer software to translate hand-drawn sketches into three dimensions. This allows models to be produced directly from the designers' digital files, giving them more control over the production process and leaving very little opportunity for misinterpretation in manufacturing.[9]

In sum, each design emerges more from Béhar's imagined narrative context than from the client's own aesthetic history—a strategy that is particularly effective when the project involves rethinking and sustaining the history of a design object that is already on the market. Today's technology-savvy consumers tend to look to stories and context to differentiate products, and the most successful brands marry the product's narrative with the tactility of the object. Although Béhar is a digitally literate designer, he remains attentive to an object's tangible effects; for him, the most important component in design is ultimately "how the bottle opens, how it feels in your hands, how it expresses an idea."[10] It is precisely this trajectory from conceptual narrative to the reality of an actual object that completes our experience of his work. On another level, it produces a framework for brand development that can be applied either to the retooling of an existing product or to the launch of a new one.

The full deployment of this methodology is demonstrated by Footprints: The Architect Collection, Béhar's 2003 and 2004 lines of shoes for Birkenstock. While preserving and rethinking signature Birkenstock features such as the natural cork and latex sole, Béhar has exploited new technologies to produce visually pleasing shoes that enhance the brand's appeal to a broader demographic. Although he built the collection around the company's central message of comfort, he has expanded the Birkenstock experience to include aesthetics. For Béhar, a successful product naturally spawns new interpretations of the original message, and this evolution can be witnessed in the Birkenstock catalogue, which now includes not only the original sandals but also Mary Janes, boots, and even gardening shoes (opposite). Béhar also created the new line's branding and logo (below).

Birkenstock Footprints: The Architect Collection, 2003, logo

The designer took his method of creating a total experience one step further with his MINI_motion collection, which merges brand identities to forge new consumer relationships and hybrid objects. When MINI approached Béhar to design a selection of car-related products, he returned with a brief that positioned it as a brand that could be sold outside car dealerships, ideally in boutiques and design shops. His MINI_motion concept revolves around the transitional act of moving in and out of a vehicle. Although the line is designed specifically for MINI, the products can be used in any car, and each object is conceived with this flexibility in mind. To produce the wide range of products, Béhar and MINI partnered with a variety of manufacturers: the shoe, for example, is made by PUMA, while the carpack is by Samsonite. Other products designed for the 2003 launch included a watch, a jacket with a convertible seat, and an experimental tent project. The watch, manufactured by the Swiss company Festina-Candino, clips onto the wrist and has a large digital face—reminiscent of the MINI's rectangular shape—whose display changes from horizontal to vertical depending on the position of the driver's hand. Revolutionizing the idea of a driving shoe, the footwear consists of two parts: a soft inner moccasin for driving slips into a treaded outer shell for walking. Perhaps the most radical of the products is the MINI_motion Environment, a tent designed to be mounted on the back of the vehicle (below). When the car is not moving, the tent creates an enclosed area that can house anything from an organic vegetable stand to a triage center. The intent behind the Environment project is to provide a multipurpose spatial framework and let the individual user program it as he or she sees fit.

MINI AID tent, a variation of the MINI_motion
Environment, 2003, rendering

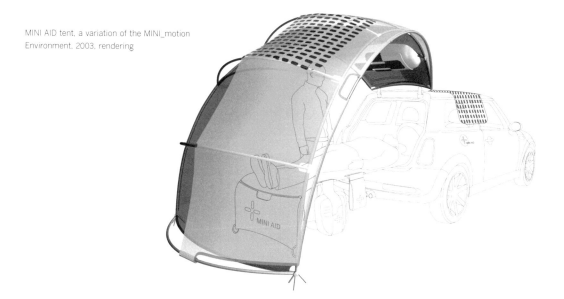

MINI's partnership with other well-known companies allows Béhar to ensure the proper execution of his designs and the quality of the finished products. It also exemplifies fuseproject's larger industrial-design philosophy, which views brand synergy as a means to develop better, more diverse products that account for user experience as well as function. Béhar recognizes that as consumer culture becomes more complex, companies are increasingly willing to share their expertise and capitalize on the widespread visual literacy of today's brand-conscious consumers.

From top: spacescent perfume bottle, 2000, rendering;
DEVO underwear packaging, 2002, rendering

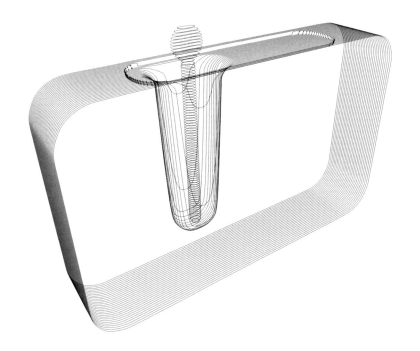

Package design plays an essential role in brand building for younger companies, as it can express the quintessential characteristics of the products they sell. With the Philou line of scented hair-care products (2000), the conceptual narrative for the bottle resulted in an asymmetrical, ergonomic form that is easy to grasp in the shower and stands out on store shelves. Many of Béhar's packaging projects are aesthetic explorations of the notion of container versus space. His 2000 design for spacescent perfume (above), produced by haasprojekt, relates the impact of the perfume to the small vial that contains it. In profile, the elegant rectangular form sinuously wraps the void, folding into a slender red cavity that appears to float in the bottle. In effect, the package provides a metanarrative of the experience of wearing perfume: the translucent volume encapsulates the product just as the scent creates an aura around the wearer. Béhar designed a similar red vial encased in translucent resin for Perfume09 (2002); the container is small enough to hold in one's palm and sturdy enough to carry in a bag.

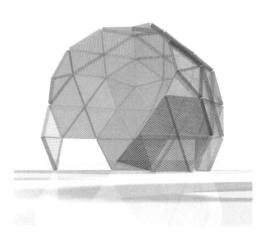

Continuing the trajectory of novel packaging is fuseproject's 2002 design for DEVO underwear (preceding page). With DEVO, Béhar devised an eco-friendly, cornstarch-based container that melts in water: Just place the entire package inside the washing machine, and the wrapper disappears during the wash cycle. Biodegradable packaging is also a primary feature of the HIP line of holistic beauty products (2003). Each cylindrical, transparent bottle is recyclable and sold in a dissolvable cornstarch box. As the product runs out, the vacuum-sealed pouch inside the bottle, which conserves the natural ingredients without preservatives, contracts to create an empty void.

Béhar inverted the premise of recognizable packaging when he designed a hooded, windbreaker-like sweater of Teflon-treated cashmere for Lutz & Patmos's fall 2001 collection (above). The technology for producing the Teflon coating is patented by DuPont, and it takes more than twelve hours to apply the treatment to the fabric. Since the waterproof coating is undetectable to the eye or the touch, Béhar communicated its function by using the sweater/windbreaker typology, underscoring this visual association with the garment's nylon storage pouch. Through clothing type, packaging, and marketing photography, Béhar was able to signal a shift in the conventional notion of what a cashmere sweater could be. He has recently begun to work with the fashion designer Hussein Chalayan, a collaboration that will almost certainly result in a variety of original product concepts.

From left: Lutz & Patmos windbreaker, 2001;
Toshiba Red Transformer, 2003;
Trium Erector, 2003, rendering

Béhar's ability to operate outside conventional notions of product design to generate novel objects reflecting our current cultural landscape can best be seen in his interpretations of technology-driven products. The wafer-thin 2003 Toshiba Red Transformer (opposite) looks simple when closed; once open, however, the screen of this sleek, multifunctional laptop slides over the keyboard to become a personal entertainment center or a flat monitor for presentations and client meetings. Béhar's conceptual narrative for Toshiba pointed out that many office laptops double as presentation devices for business travel. By morphing the laptop, presentation monitor, and entertainment unit into one multifunctional product, he generated an essential tool that addresses the needs of today's consumer. Similarly, the Aliph Jawbone audio headset (2003) is a simple design that clips onto the ear and seems more like jewelry than a technological tool. The ergonomic headset filters out all ancillary sound, rendering audible only the voices in the user's conversation.

Béhar has recently expanded his design practice to include furniture, lighting, and speculative housing prototypes as well as conceptual and retail spaces. His Trium Erector (opposite), a 2003 housing-system prototype, resembles both a large-scale Lego construction and a deconstructed geodesic dome. The system relies on serially employed triangular panels that can be assembled in a variety of configurations to form a self-supporting, monocoque-like structure. The colorful modules are designed to be produced with 3D-TEX, a woven, dimpled polyester fabric that Béhar proposes coating with a light resin, enabling it to become a translucent structural element.[11]

While the Trium Erector is highly speculative, it encapsulates the singular qualities of Béhar's narrative-inspired approach to design, a process that offers a fresh and viable alternative to his profession's customary problem-solving methodology. Equally comfortable in the realms of lifestyle objects and brand development, he is one of the few industrial designers whose ability truly spans the breadth and plumbs the uncharted depths of the discipline. Moreover, his talent for orchestrating corporate synergies reflects a radical new role for the industrial designer—that of a cultural critic producing conceptual ideas, generating new typologies, and fusing brand relationships that will transform the course of industrial design in the twenty-first century.

13

notes

[1] Yves Béhar, interview by author, July 14, 2003. [2] For more on the "good design" syndrome, see Arthur J. Pulos, *The American Design Adventure: 1940–1975* (Cambridge, MA: MIT Press, 1988), 110–21, and Jocelyn de Noblet, ed., *Industrial Design: Reflection of a Century* (Paris: Flammarion / APCI, 1993). More information on the Bauhaus and the Ulm Design School may be found in Hans M. Wingler, *The Bauhaus* (Cambridge, MA: MIT Press, 1990) and Herbert Lindinger, ed., *Ulm Design: The Morality of Objects, Hochschule für Gestaltung Ulm, 1953–1968* (Cambridge, MA: MIT Press, 1990). Also see Gillian Naylor, *The Bauhaus Reassessed: Sources and Design Theory* (New York: E. P. Dutton, 1985). [3] Béhar, interview by author, July 21, 2003. [4] Béhar, interview by author, July 14, 2003. [5] Ibid. [6] Béhar, interview by author, August 3, 2003. [7] See K. Michael Hays, ed., *Architecture Theory Since 1968* (Cambridge, MA: MIT Press / New York: Columbia Books of Architecture, 1998), x–xv. [8] Béhar, interview by author, July 14, 2003. [9] Béhar, interview by author, August 3, 2003. [10] Béhar, interview by author, July 21, 2003. [11] Béhar's Trium Erector was one of three projects commissioned by *I.D.* magazine to explore new uses for 3D-TEX. See Julie Lasky, "Material Migrations," *I.D.* 50, no. 7 (November 2003): 66–69.

projects

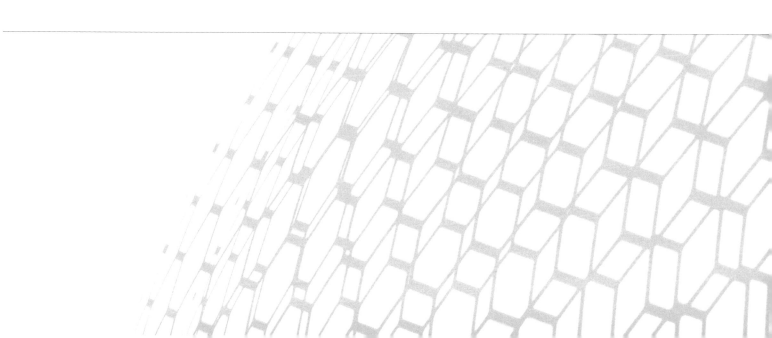

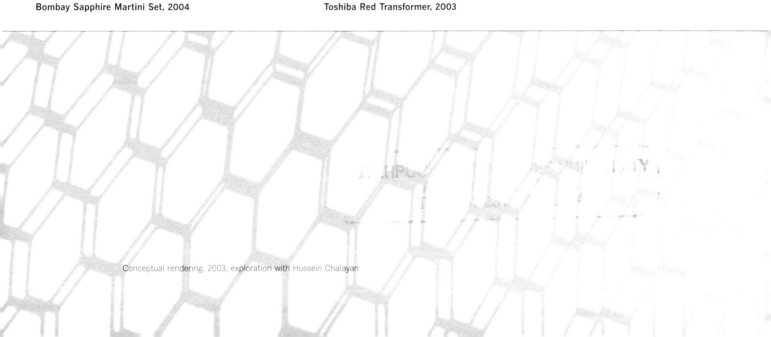

Conceptual rendering, 2003, exploration with Hussein Chalayan

MINI_motion Carpack, 2003
MINI_motion Environment, 2003
MINI_motion Jacket with Convertible Seat, 2003
MINI_motion Two-Part Driving Shoe, 2003
MINI_motion Watch, 2003

move

With the success of the new MINI, released in 2001 as a retooling of the classic 1959 model, Americans have finally embraced a European sensibility in their relationship to their cars. The MINI is a lightly worn shell, suitable for the urban crunch as well as the open road. Small and agile, it parks well, maneuvers tight corners, and, as the shortest car currently available on the American market, is unlikely to intimidate pedestrians.

In designing the MINI_motion product line, Béhar recognized that the vehicle's suitability for the urban lifestyle is more than a matter of size. The car has become one of many components of the commuter's daily journey, and commuting time just a small part of an always-mobile, always-connected, twenty-four-hour day. The MINI_motion collection helps drivers to transition from the car to any other part of their lives without, so to speak, shifting gears.

The first products in the MINI_motion line, each of which incorporates a rounded-rectangle motif that echoes the car's contours, are all indebted to the design tradition of hybridity: objects that take various forms in order to serve diverse functions. The two-part shoe, manufactured by PUMA, combines a soft, flexible inner moccasin, ideal for driving, with a stronger, treaded exterior

Opposite: MINI_motion motif, 2003

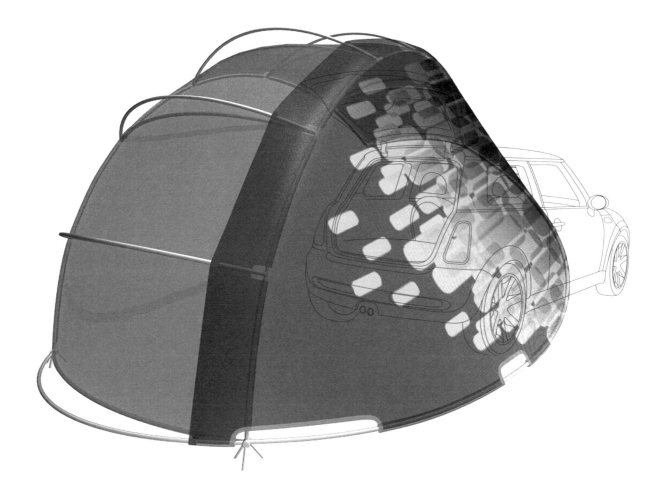

that the driver dons when exiting the car for the street. The Gore-Tex jacket integrates map pockets, a small detachable bag, and an extension that flips down in the back to become a portable seat. The watch has an oversize digital display that switches from horizontal to vertical depending on the position of the driver's arm. The carpack, manufactured by Samsonite, attaches to the seat belt for easy accessibility and doubles as a pedestrian messenger bag.

Béhar's articulation of the relationship between design and narrative is evident in his designs for the MINI. As these products expand their functionality beyond the automobile, they also expand their psychological presence in the user's daily routine, turning a single consumer choice—the purchase of a car—into an integrated lifestyle.

Top, from left: MINI_motion watch, two-part driving shoe, carpack, jacket, 2003
Bottom: MINI_motion Environment, 2003, rendering

Birkenstock Footprints: The Architect Collection, 2003, insole rendering

Learning Shoe, 2000
Birkenstock Footprints: The Architect Collection, 2003, 2004
Birki Garden Shoe, 2004

step

Birkenstock has been making sensible shoes since 1774. Economical, eco-friendly, and above all ergonomic, the German brand found a ready audience among younger buyers when it was introduced to the U.S. market in the late 1960s. Birkenstock's quirky but comfortable aesthetic was a big hit with hippies and other counterculture types who embraced the shoe, with its humanist values, as an icon of their lifestyle.

Today these values have migrated into the mainstream, and consumers looking for recyclable products and "green" design also demand a degree of style consciousness and aesthetic sophistication. Birkenstock, seeking to update its brand identity with an infusion of urban chic, hired fuseproject to create a new line of shoes.

Béhar came to Birkenstock's attention in 2000, when two of his concept-shoe designs were included in SFMOMA's exhibition *Design Afoot: Athletic Shoes, 1995–2000*. The concepts—which included a high-tech shoe that "learns" a particular wearer's foot and can be used to customize future pairs—seemed like a natural fit for Birkenstock's footwear philosophy.

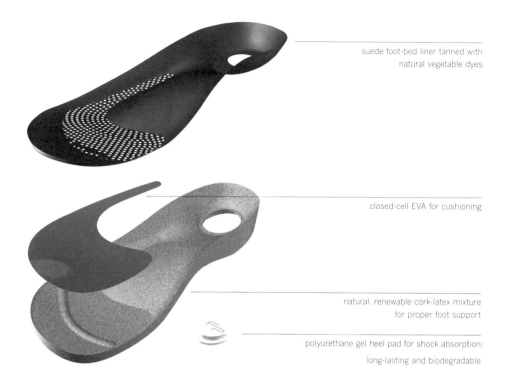

suede foot-bed liner tanned with
natural vegetable dyes

closed-cell EVA for cushioning

natural, renewable cork-latex mixture
for proper foot support

polyurethane gel heel pad for shock absorption:
long-lasting and biodegradable

The 2003 and 2004 collections incorporate a number of classic Birkenstock features, such as wide toe boxes and cork foot beds, but they also enhance the shoes' environmentally friendly design with new technology. The heels feature biodegradable gel padding, and the soles and toe supports are layered with recyclable foam. The outsoles and foot beds are replaceable, so the long life span of the shoes—an essential Birkenstock feature—can be even further extended.

The name of the line, Footprints: The Architect Collection, alludes to the shoes' design as sophisticated housing for the feet. Softly streamlined forms respect the foot's contours while showing off the sculptural beauty of the natural, though technologically enhanced, materials. One of the collection's more whimsical features—acrylic windows in the soles that make the cork inlays visible from the outside—reminds wearers that at the heart of the new design, the shoe is still a classic Birkenstock.

Birkenstock Footprints: The Architect Collection, 2003, diagram of insole construction

Birkenstock Footprints: The Architect Collection, 2003.
Clockwise from left: Vigo, Zamora, Avila

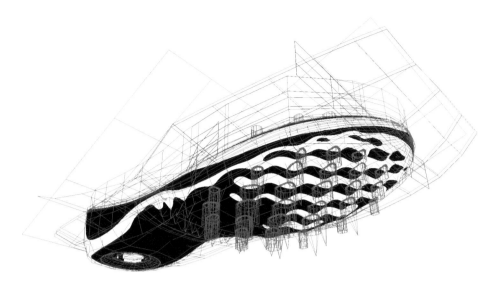

Birkenstock Footprints: The Architect Collection, 2004,
rendering and view of outsole
Opposite: Learning Shoe, 2000, concept

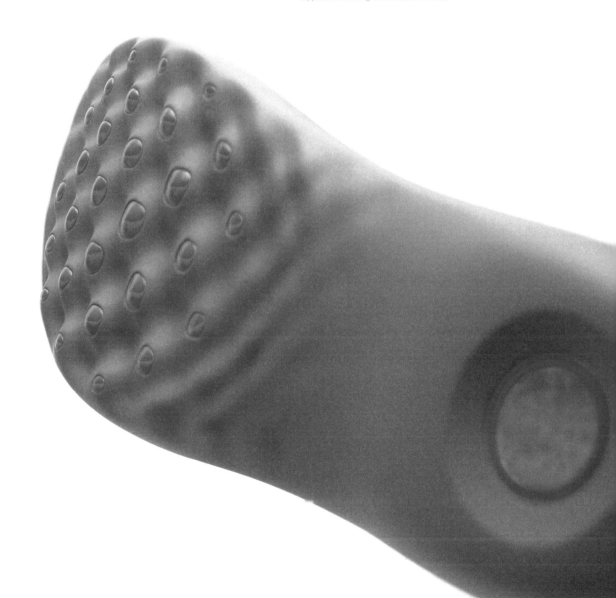

Lutz & Patmos windbreaker, 2001, detail

touch

Cashmere is the quintessential luxury item: It is expensive, it requires careful handling, and as a simple defense against the cold it may seem needlessly indulgent. But the plush fabric has an irresistible quality, and it never goes out of style.

The universal visual and tactile appeal of cashmere is what brought Béhar together with the New York fashion house Lutz & Patmos. Every year Tina Lutz and Marcia Patmos, who specialize in cashmere clothing, commission an artist or designer from outside the fashion world to contribute to their collection. For the fall 2001 line they selected Béhar, an obvious choice given his experience with a wide range of design genres and his interest in experimenting with traditional materials.

Béhar's approach, in a strategy typical of fuseproject, was to infuse the organic with the technological, transforming this usually vulnerable fabric into something more resilient and suitable for everyday outdoor wear. In a process patented by DuPont, the cashmere is treated while it is still at the mill. Instead of spraying the surface of the finished garment, the raw wool is soaked in a Teflon bath that coats the individual fibers and renders them water- and oil-resistant. Because it is applied before the cashmere is woven, the treatment does not affect the fabric's look or feel; it has the same drape, texture, and fit as untreated cashmere. The hooded sweater is thus transformed into a waterproof windbreaker.

Béhar's design concept for the windbreaker also includes a compact and translucent nylon storage pouch, which not only protects the cashmere from humidity and moths, but also visually reinforces the sweater's versatility and practical appeal.

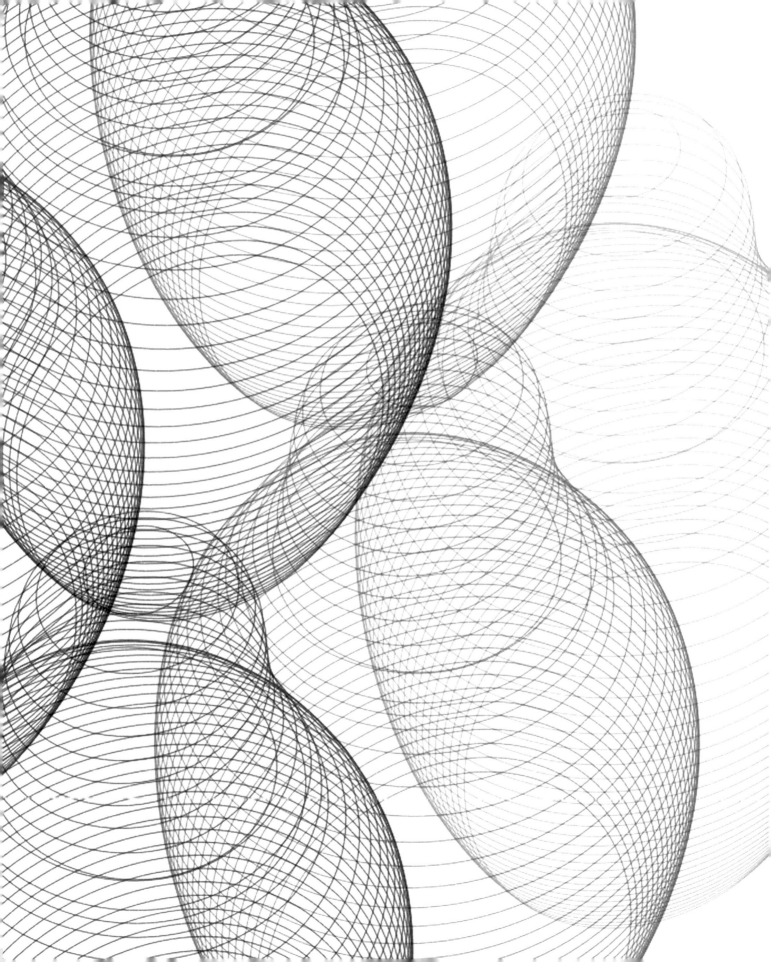

hold

Whether a teapot, vodka decanter, or perfume bottle, a simple container is one of the most basic and iconic forms a designer can create. In its many concepts for bottles and packaging, fuseproject has explored the essential qualities of "containment" as both an abstract aesthetic and a particular design challenge.

The bottle Béhar designed in 2000 for spacescent perfume is an exercise in negative space. The perfume itself, enclosed in a classic vial-shaped chamber, is surrounded by a brick of clear resin bordered with a slim band of red. In spite of its transparency, the brick exaggerates and emphasizes the protected zone around the precious fluid. The packaging for Perfume09 is similar, with the portable vial encased in rubber and packaged in foam, making it impervious to the perils of travel in a suitcase, purse, or pocket.

Béhar's 2000 design for Philou's selection of teen hair-care products turned a simple, eccentric shape into an icon. The teardrop form, based on a tilted oval, is evocative of sensuality without being overtly sexual. Ergonomic and softly textured, its surface is not only seductively tactile and visually appealing, but also makes the bottle easy to grasp in the shower.

Opposite: Philou shampoo and conditioner bottles, 2000, rendering

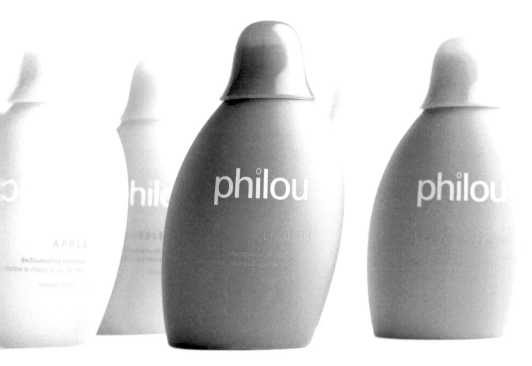

For the HIP line of French beauty products, fuseproject designed an innovative packaging system in 2003. HIP advertises its cosmetics as organic, eco-friendly, and pure, using a color-coded system of bright hues to distinguish one product line from another. Fuseproject's design encloses the product in a vacuum-sealed pouch within a translucent, streamlined bottle. The substance stays fresh, protected, and uncontaminated, while the clear exterior allows users to see how much of the product they have left.

HIP bottles are packaged in biodegradable, cornstarch-based boxes that dissolve in water instead of ending up as landfill. Fuseproject used the same technology to create wrappers for DEVO underwear in 2002; a brand-new package of DEVO briefs can be tossed, unopened, directly into the washing machine.

In addition to using unconventional materials, these designs are playful and experimental. They achieve their aesthetic power by exploring the opposing parameters that define all containers: inside versus outside, the visual versus the tactile, accessibility versus protectiveness.

Philou shampoo bottles, 2000
Opposite: HIP bottles, 2003

/arc-en-ciel

HIP

HOLISPA

holi-gloss

ONCTION CORPS
allège, tonifie, stimule le désir

BODY UNCTION
to lighten, tone up and sex appeal

/arc-en-ciel

HIP

HOLISPA

holi-gloss

ONCTION CORPS
hydrate, illumine, anti-âge

BODY UNCTION
to moisturise and enhance, anti-aging

/arc-en-ciel

HIP

HOLISPA

holi-gloss

/arc-en-ciel

HIP

HOLISPA

holi-gloss

/arc-en-ciel

HIP

HOLISPA

holi-gloss

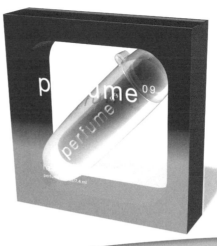

From top: Perfume09 bottle and packaging, 2002, rendering;
spacescent perfume bottle, 2000, rendering
Opposite: Bombay Sapphire martini set, 2004

space scent 001

Avo Cell Phone, 1999

HP Pavilion Computer, 2000

PeoplePC Computer Accessories, 2000

Aliph Jawbone Audio Headset, 2003

MovieBeam Receiver, 2003

Toshiba Red Transformer, 2003

connect

In the realm of high-tech gadgetry, good design is usually measured in terms of technological muscle: power, speed, and accuracy. What is often lost in this equation is the human factor—the psychological relationship between users and their devices. This relationship, based more on comfort, familiarity, and accessibility than raw technological might, is an intangible that many programmers and engineers choose to ignore.

Fuseproject's designs for computers, mobile phones, and other tech gear take the human component as their principal inspiration. In 2000, while creating a design language for a new line of HP computers, the studio developed an aesthetic that was inviting rather than astonishing: soft materials, a smaller scale, and compact, straightforward engineering features. Likewise, a line of accessories for PeoplePC computers, also from 2000, emphasizes simple practicality and personalization. The rounded forms and whimsical look of the keyboard, monitor organizer, and other components allow computer users to customize, and thus humanize, their work environments.

Opposite: HP Pavilion computer, 2000

hp *pavilion* 2000

The 2003 Toshiba Red Transformer, a seventeen-inch laptop with red lacquer finish, is designed for adaptability. Using a complex hinge, the computer screen converts from a laptop configuration into a presentation display or a viewing screen for personal entertainment. Customization is also the key feature of Avo, Béhar's 1999 cell phone design. The phone's interchangeable exterior, made of a soft, pliable plastic, molds to fit the user's hand and, over time, will retain that individualized shape. The 2003 wireless Aliph Jawbone headset combines a stylish design with technology that suppresses ambient noise, allowing the user to comfortably filter out distractions.

In all of these products, innovative design serves as an interface between technology and people, offering more choices and increased physical comfort. Technology is treated as a servant—a tool that adapts to our needs rather than one that requires our submission.

From left: PeoplePC mouse pad, 2000; Avo cell phone, 1999; MovieBeam receiver, 2003
Opposite: Aliph Jawbone audio headset, 2003

Toshiba Red Transformer, 2003

expand

Furniture can be considered a kind of technology, allowing the body to extend the parameters of its natural abilities and adapt to different environments. Emphasizing flexibility and hybrid forms, fuseproject's designs for furniture and architectural spaces articulate this ideal of expansion and adaptability.

The 2001 Opium Mat is a portable piece of furniture that functions as a daybed or movable lounge chair. Based on the tatami mat, a woven straw floor covering in traditional Japanese homes, it is constructed from laser-cut cedar slats held together by a backbone of cable springs. The collapsible cable system allows the mat to be stored flat or rolled. The 2003 Herman Miller lamp is another highly elastic design. It unfolds elegantly out of a slim cylindrical form that makes a compact footprint on the desk or floor. The lamp twists and rotates to throw illumination in any direction, providing either task or ambient lighting.

Opposite: Opium Mat lounge chair, 2001, detail

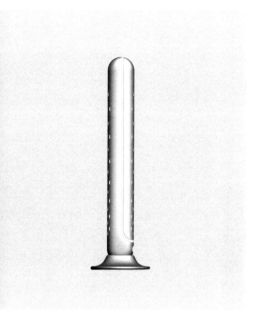

The Dream Room, a conceptual space originally created for *Dwell* magazine in 2001, is enclosed by wall-size LCD displays. The "digital wallpaper" can be programmed to mimic ordinary walls decorated with hanging pictures, or it can project floor-to-ceiling scenes of the outdoors or other vistas, creating the illusion of a landscape that extends beyond the room. This technology works much like a television or computer screen to blur the boundaries of the space both visually and psychologically.

In 2003 fuseproject ventured into built architecture with the Friend concept store, a mixed-use space in San Francisco that combines a retail store with an art gallery and a community gathering place. This flexible, hybrid environment, in which design objects are both exhibited and sold, is itself an architectural prototype intended to encourage public dialogue about design.

Herman Miller task lighting, 2003, renderings
Opposite: Friend concept store, 2003

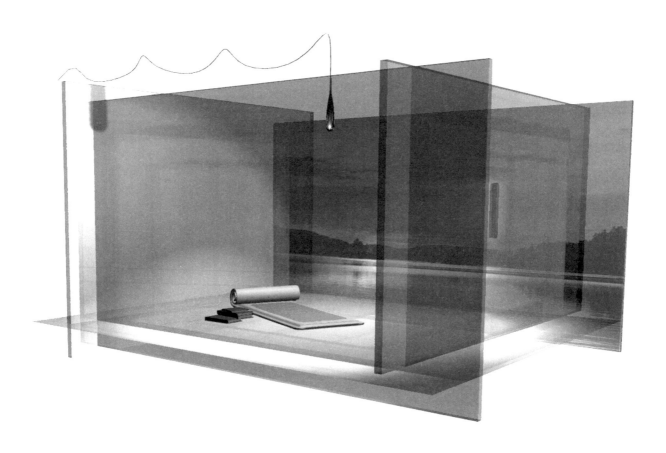

Dream Room, 2001, rendering
Opposite: Trium Erector, 2003, rendering

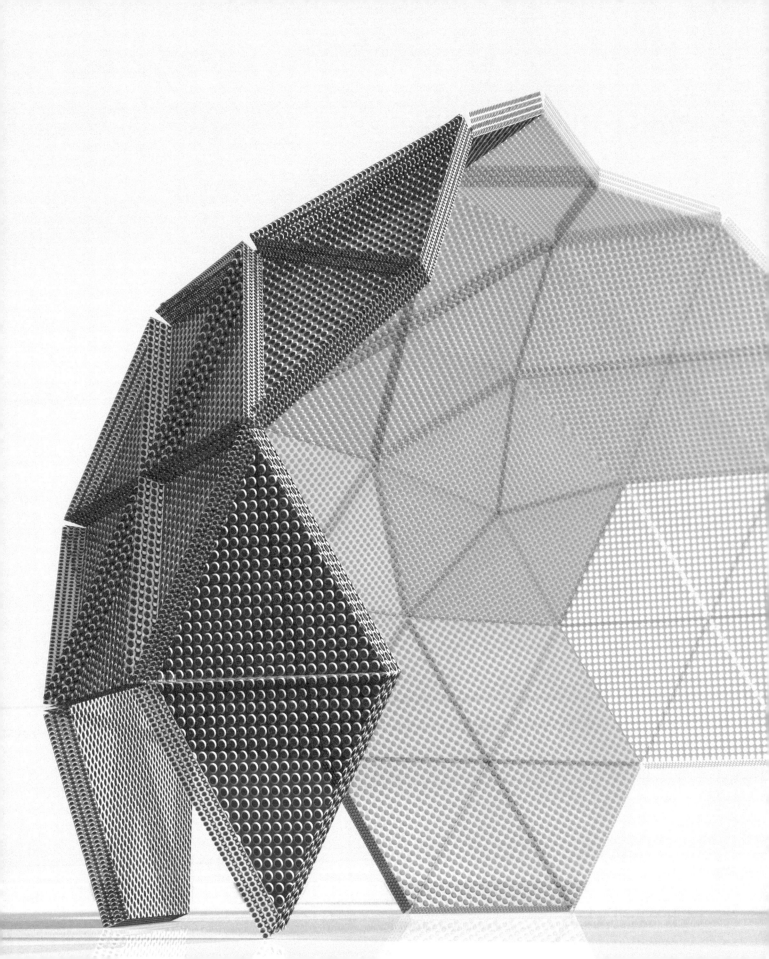

project information

move

MINI_motion Carpack, 2003
Materials: EVA, molded EVA nylon
Client: MINI USA/BMW Group
Manufacturer: Samsonite Corporation
Design team: Yves Béhar with Johan Liden
Status: commercially available

MINI_motion Environment, 2003
Materials: stretch nylon, aluminum tubing, milled aluminum
Client: MINI USA/BMW Group
Manufacturer: Moss
Design team: Yves Béhar with Geoffrey Petrizzi and
Shawn Sinyork
Status: experimental project

MINI_motion Jacket with Convertible Seat, 2003
Materials: Gore-Tex, synthetic materials
Client/manufacturer: MINI USA/BMW Group
Design team: Yves Béhar with Johan Liden
Status: commercially available

MINI_motion Two-Part Driving Shoe, 2003
Materials: TPU, nylon
Client: MINI USA/BMW Group
Manufacturer: PUMA
Design team: Yves Béhar with Shawn Sinyork and
Eskil Tomozy
Status: commercially available

MINI_motion Watch, 2003
Materials: TPU, stainless steel
Client: MINI USA/BMW Group
Manufacturer: Festina-Candino Watch Ltd.
Design team: Yves Béhar with Shawn Sinyork and
Eskil Tomozy
Status: commercially available

step

Learning Shoe, 2000
Materials: injection-molded PVC
Client: San Francisco Museum of Modern Art
Design team: Yves Béhar with Eskil Tomozy and
City Electric
Status: concept

Birkenstock Footprints: The Architect Collection, 2003
Materials: suede, leather, recycled cork, natural rubber,
vegetable dyes, cork-latex blend, biodegradable TPU,
biodegradable polyurethane gel
Client/manufacturer: Birkenstock USA
Design team: Yves Béhar with Johan Liden and
Geoffrey Petrizzi
Status: commercially available

Birkenstock Footprints: The Architect Collection, 2004
Materials: suede, leather, recycled cork, natural rubber,
vegetable dyes, cork-latex blend, biodegradable TPU,
biodegradable polyurethane gel
Client/manufacturer: Birkenstock USA
Design team: Yves Béhar with Geoffrey Petrizzi and
Pichaya Puttorngul
Status: commercially available 2004–05

Birki Garden Shoe, 2004
Materials: biodegradable TPU, EVA
Client/manufacturer: Birkenstock USA
Design team: Yves Béhar with Shawn Sinyork,
Geoffrey Petrizzi, and Johan Liden
Status: commercially available

touch

Lutz & Patmos Windbreaker, 2001
Materials: Teflon-coated cashmere
Client/manufacturer: Lutz & Patmos Inc.
Design team: Yves Béhar with Tina Lutz
Status: commercially available

hold

Philou Shampoo and Conditioner Bottles, 2000
Materials: low-density polyethylene, injection-molded ABS,
coated polyethylene
Client/manufacturer: Philou
Designer: Yves Béhar
Status: commercially available

spacescent Perfume Bottle and Packaging, 2000
Materials: polyurethane resin, enamel, rubber
Client: haasprojekt/Brent Haas
Manufacturer: Corey Jones
Design team: Yves Béhar with Johan Liden
Status: commercially available

DEVO Underwear Packaging, 2002
Materials: cornstarch
Client: DEVO
Design team: Yves Béhar with Johan Liden and Angelita Tadeo
Status: prototype

Perfume09 Bottle and Packaging, 2002
Materials: polyurethane rubber, glass, EVA foam
Client/manufacturer: haasprojekt/Brent Haas
Design team: Yves Béhar with Johan Liden and
Geoffrey Petrizzi
Status: commercially available

HIP Bottles and Packaging, 2003
Materials: recyclable polypropylene, cornstarch
Client/manufacturer: HIP
Design team: Yves Béhar with Johan Liden
Status: commercially available

Bombay Sapphire Martini Set, 2004
Materials: glass
Client/manufacturer: Bombay Spirits Company U.S.A.
Design team: Yves Béhar with Pichaya Puttorngul and
Eskil Tomozy
Status: commercially available fall 2004

connect

Avo Cell Phone, 1999
Materials: injection-molded ABS, rubber
Client: frog design inc.
Design team (at frog design): Yves Béhar with
Joshua Morenstein
Status: prototype

HP Pavilion Computer, 2000
Materials: injection-molded ABS
Client/manufacturer: Hewlett-Packard Company
Design team: Yves Béhar with Joshua Morenstein,
Peter Lee (HP), and Gecko
Status: commercially available

PeoplePC Computer Accessories, 2000
Materials: nylon EVA, Velcro, injection-molded ABS
Client/manufacturer: PeoplePC Inc.
Design team: Yves Béhar with Joshua Morenstein, City
Electric, and Gecko
Status: commercially available

Aliph Jawbone Audio Headset, 2003
Materials: injection-molded ABS, aluminum
Client/manufacturer: Aliph
Design team: Yves Béhar with Geoffrey Petrizzi,
Youjin Nam, and Johan Liden
Status: commercially available

MovieBeam Receiver, 2003
Materials: injection-molded ABS
Client: MovieBeam, a Walt Disney Co., and Samsung
Manufacturer: Samsung
Design team: Yves Béhar with Johan Liden, Geoffrey
Petrizzi, Aldis Rauda, Shawn Sinyork, and Richard
Drysdale (MovieBeam)
Status: commercially available

Toshiba Red Transformer, 2003
Materials: injection-molded ABS
Client: Toshiba America Inc.
Design team: Yves Béhar with Geoffrey Petrizzi,
Shawn Sinyork, and Pichaya Puttorngul
Status: prototype

expand

Dream Room, 2001
Materials: liquid crystal displays
Client: *Dwell* magazine
Design team: Yves Béhar with Lisa Lo and
Geoffrey Petrizzi
Status: concept

Opium Mat Lounge Chair, 2001
Materials: yellow cedar, steel cable, machined aluminum
Design team: Yves Béhar with Johan Liden and
Geoffrey Petrizzi
Status: prototype

Friend Concept Store, 2003
Location: San Francisco
Size: 1,500 square feet
Materials: salvaged white oak, recycled tires, acrylic
Client: Friend
Design team: Yves Béhar with Johan Liden and
Franz Schnaas
Architect: Todd Davis
Status: completed

Herman Miller Task Lighting, 2003
Materials: ABS, stamped aluminum
Client: Herman Miller Inc.
Design team: Yves Béhar with Johan Liden,
Eskil Tomozy, and Shawn Sinyork
Status: prototype

Trium Erector, 2003
Materials: 3D-TEX (resin-coated polyester and Lycra fabric)
Client: *I.D.* magazine
Design team: Yves Béhar with Eskil Tomozy
Status: concept

selected bibliography

Anargyros, Sophie Tasma. "Yves Béhar, une vision rêvée" [Yves Béhar, a dreamlike vision]. *Intramuros* 108 (August–September 2003): 44–49, 107.

Béhar, Yves. "PeoplePC Accessories, Functional Fashion." *Innovation* 21, no. 3 (Fall 2002): 69.

Cameron, Kristi. "Peep Shoe." *Metropolis* 22, no. 11 (July 2003): 40.

"Full Throttle." *Surface* 41 (2003): 110.

"Fuseproject." *Zoo Quarterly* 12 (March 2002): 174–77.

Galway, Katrina. "On the Scent of Product Lust." *Innovation* 20, no. 4 (Winter 2001): 50.

Heller, Steven, and Yves Béhar. "Improving Everyday Life." *The Education of a Design Entrepreneur,* edited by Steven Heller. New York: Allworth Press, 2002.

Hirasuna, Delphine. "From Hippie to Hip: Birkenstock Goes Urban." *@issue* 9, no. 1 (Fall 2003): 6–11.

Klausner, Amos. "Slipping into Smart Fabrics." *Core77,* http://www.core77.com/materials_processes/art_smartfab.asp.

Labonte, Adrienne Marie. "All About Yves." *7x7* (February–March 2002): 32.

Lasky, Julie. "Material Migrations." *I.D.* 50, no. 7 (November 2003): 66–69.

Lupton, Ellen, ed. *Inside Design Now: National Design Triennial.* New York: Cooper-Hewitt, National Design Museum / Princeton Architectural Press, 2003.

———. *Skin: Surface, Substance, and Design.* New York: Cooper-Hewitt, National Design Museum / Princeton Architectural Press, 2002.

"Mix Master—Yves Béhar, Global Designer." *City* 27 (Winter 2003): 48–51.

Nunn, Jennie. "Heeling Power." *Surface* 43 (2003): 90.

Nussbaum, Bruce. "Annual Design Awards: Elegance Meets Common Sense." *BusinessWeek* (July 7, 2003): 72–73.

O'Loughlin, Sandra. "Hip Shoes, Not for Hippies." *Brandweek* (November 3, 2003): 18–20.

Patton, Phil. "Mini Fever Spawns Line of Accessories." *New York Times* (March 6, 2003).

Said, Carolyn. "More than Meets the Eye." *San Francisco Chronicle* (July 2, 2003).

———. "A New Line for Birkenstock." *San Francisco Chronicle* (July 2, 2003).

Sardar, Zahid. "Industrial Strength." *Interior Design* 73, no. 3 (March 2002): 191–93.

———. "Mind over Matter." *San Francisco Chronicle Magazine* (February 17, 2002): 28.

Stang, Alanna. "Mini Me." *I.D.* 50, no. 6 (September–October 2003): 20.

Stein, Jenny. "Sunshine State." *Surface* 32 (2001): 78–80.

Wilson, Eric. "The Month in Fashion." *W* 32, no. 4 (April 2003): 76.

Wohlfarth, Jenny, ed. "47th Annual Design Review." *I.D.* 48, no. 5 (August 2001): 96–111, 162–87.

Wolff, Laetitia, and Wang Xu, eds. *Yves Béhar: fuseproject.* Design Focus. Corte Madera, CA: Gingko Press; Beijing: China Youth Press, 2001.